THERMOGRAPHY:
A NEW WAY TO PROFIT FROM THE ENERGY CRISIS

THERMOGRAPHY:
A New Way to Profit from the Energy Crisis

Richard H. Munis

EXPOSITION PRESS
HICKSVILLE, NEW YORK

First Edition

ISBN 0-682-48462-8

Printed in the United States of America

CONTENTS

ACKNOWLEDGMENTS

I wish to express my gratitude to Mr. Dietrich Baeu of the AGA Corporation for providing the thermograms and photographs of the AGA Thermovision infrared camera system.

THERMOGRAPHY: A NEW WAY TO PROFIT FROM THE ENERGY CRISIS

1

PROBLEMS PRESENTED BY THE ENERGY CRISIS

The top of the sign reads: Premium 70c per gallon. At the bottom: Sorry, no gas today. You rub your eyes in disbelief. A few months ago the same premium gasoline was available at about 45c per gallon, and you could buy it anytime and anywhere that you happened to stop to fill your tank. You are still somewhat stunned as you arrive home without having found a gas station that could sell you enough gasoline to fill your tank. Upon opening the mail that arrived today, you find a bill for the last refill of heating oil. Again you stare in disbelief at the figure of 44c per gallon that is written on the bill. Why, just last winter you were paying only 21c per gallon. As you stand there staring at your heating bill, you have mixed feelings of frustration and anger.

What is happening in the United States today is and will be happening in every industrialized nation in the world for the foreseeable future. Everyone has an opinion, but nobody seems to know exactly why there is a shortage of energy and why the cost of that energy has risen so sharply during the past year. Some blame the so-called energy crisis on the Arabs, some blame the oil companies, while others blame everyone they can for being wasteful of energy. Whatever and whoever is to blame for our predicament today, there is one fact of which you can be absolutely sure: The cost of energy will continue to rise over the long term. If you do not believe that statement, just tune in to your radio or read the latest news.

Generally speaking, how has the average homeowner reacted

to this drastic change in energy cost and availability? The most obvious action is to turn down the thermostat a few degrees and wear heavier clothes around the house. Those fortunate people who have fireplaces and wood stoves in their homes have been able to compensate somewhat for the lowered thermostat by heating with wood; however, unless they own land from which they can cut their own firewood, their costs have not been substantially reduced, since the cost of firewood also has risen sharply during the past year.

Once the average homeowner has lowered the thermostat and/or bought a wood stove and a few cords of firewood, his next thought is to look at the windows and doors of his home to see if he can lower heating costs by putting plastic on all the windows and installing storm doors and weather stripping around doors. After having finished that much, he is usually convinced that he has done all that can be done and is content for the present (or until the price of heating oil takes another large jump).

Now that we have seen the reaction of the average homeowner, let us examine how the average businessman is affected. As the cost of energy rises sharply the most natural action for an average businessman is to cut production. When he shuts down production lines, he must lay off workers and increase substantially the prices of existing products to help pay the increased costs (direct and indirect) of energy. Finally, if the situation gets desperate enough, he is forced to declare bankruptcy; and the business closes its doors.

In the meantime the problems of the homeowner and businessman have just begun. Because the homeowner has turned down the thermostat and stopped some heat loss with storm doors and weather stripping, and because some industrial production lines are being shut down, less electricity is being used. And since less electricity is being consumed, the utilities are losing money because they now have a large amount of peak generating capacity that is not being used. In order to continue to operate at a profit the utilities must pass their increased cost of oil to the consumer.

By now the reader is asking how this sequence of events which began in the winter of 1974-1975 can be turned into a profit for him. The next three chapters lay the foundation for an explanation of an exciting new technology that promises to become, within the next five years, the single most important technology for effecting conservation of energy in each and every building across the United States.

The purpose of this book is to introduce businessmen and potential businessmen to the new and exciting technology of building heat loss inspection. With the availability of energy becoming less and less certain and the higher cost of this energy becoming more and more certain, for the consumer absolutely no other alternative exists (solar, nuclear, and wind energy notwithstanding) to increasing the amount and efficiency of insulation in each and every building in this country. There is a potential for a significant amount of money to be made if the reader takes action now to get in on the ground floor of this new industry.

2

THERMOGRAPHY—A NEW TECHNIQUE FOR LOCATING BUILDING HEAT LOSSES

In the last twenty-five years the world has witnessed an explosion of technological development unparalleled in the history of the world. It has become rather routine to turn on a television set to watch astronauts land on the moon, or to watch events occurring half way around the world which are reaching the United States via satellites orbiting the earth. Likewise, we read and hear about laser-guided "smart bombs" being used by our military, and we attempt to understand how man has developed the capability to destroy civilization with nuclear weapons.

Aside from these spectacular developments, many more less spectacular technological breakthroughs are proving to be extremely useful to our society. The most important of these technological advances, for the purpose of this book, is that of the detection of infrared energy (heat) radiated from an object. Since this type of energy is invisible to the human eye, it went undetected until the advent of special devices that could "see" the emission of heat from objects.

Within the last twenty-five years, technological development has brought the science of heat detection to the threshold of several practical applications. The most significant development in this respect has been infrared thermography.

Infrared thermography uses an infrared camera system to detect heat radiated from an object. The heat is converted into an electrical signal by an infrared scanner, or camera, a device not unlike a regular camera in that its basic function is to change one form of energy into another.

After the heat is converted into an electrical signal, this signal is displayed on a TV-type screen as a "heat picture" of the object. This picture is not as distinct as a regular photograph and is composed essentially of a gray scale which has continuous contrast ranging from black to white. The areas of higher temperature appear lighter; the areas of lower temperature are darker. Intermediate shades of gray indicate variations between the extremes of temperature. For example, if the infrared camera is looking at a warm or hot object, the heat picture will be on the light side of gray. If the camera is looking at a relatively cool object, the heat picture will be on the dark side of gray.

In practice, the heat picture seen on the TV-type screen can be photographed with a Polaroid or 35-mm camera. The resulting photograph of the heat picture displayed on the TV-type screen is called a thermogram. This thermogram gives complete information on the surface temperature of an object. Any object, such as the wall of a building, can be inspected with an infrared camera system to determine what locations on the wall are warmer, or cooler, than other locations.

It should be emphasized that an infrared camera system can detect only temperature differences across the surface of an object. The system itself will not give any information on how much heat is being lost; however, it is possible to compute approximate heat loss rates from any thermogram if certain procedures are followed.

Fig. 1 shows a closeup view of the AGA Thermovision infrared camera system. The unit on the left is the infrared camera. The function of the lens on the front of this camera is much like that of a lens on a regular camera, except that it forms an image of heat on a special heat sensing device instead of an image of light on a film. The heat sensing detector is located toward the rear of the infrared camera and must be cooled with ¼ pint of low temperature coolant (liquid nitrogen). The container for the

FIG. 1. *AGA Thermovision infrared camera system*

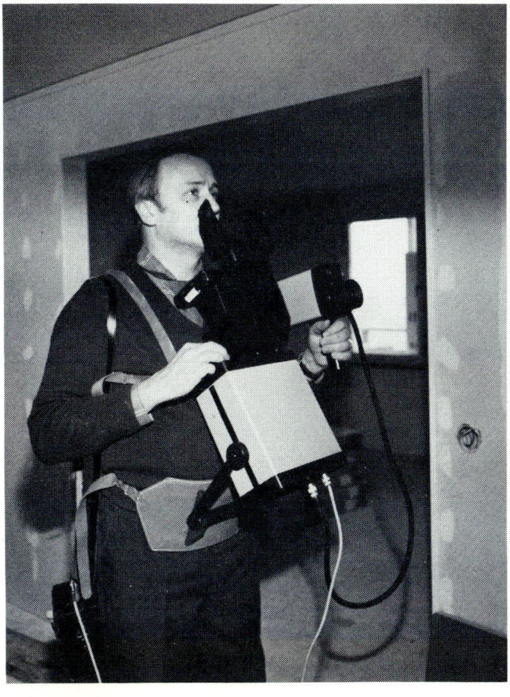

FIG. 2. *AGA Thermovision system in use*

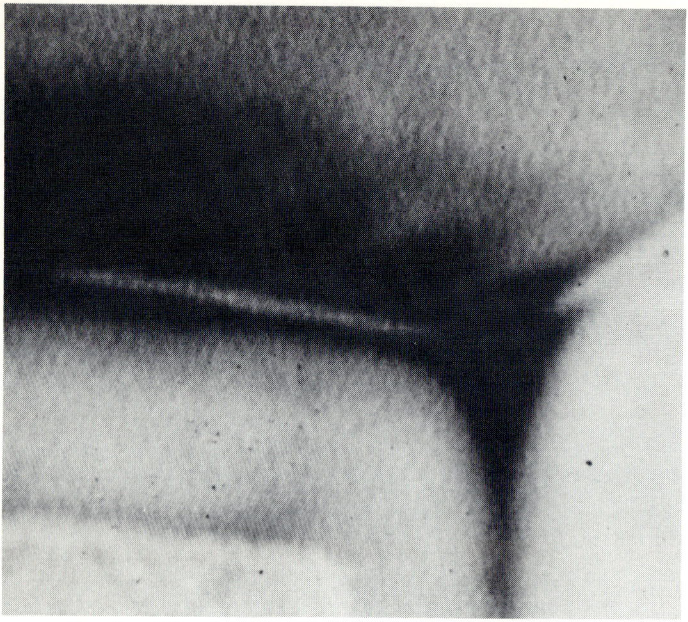

FIG. 3. *Thermogram of junction of walls and ceiling*

FIG. 4. *Thermogram of exterior of house*

coolant is located at the rear of the infrared camera, and the cap for the container can be seen on the top of the camera case. The unit on the right contains the TV-type screen on which the heat picture is presented and can be photographed by a camera attachment. The weight of the infrared camera is slightly more than three pounds, and that of the TV-type display unit is approximately ten pounds.

Fig. 2 shows the AGA Thermovision infrared camera system being used to inspect the interior of a new building for heat leaks during construction. Notice that the infrared camera is being held in the left hand, while the TV-type display unit is carried by a special chest harness. Also visible in this photograph is the Polaroid camera attached to the TV-type display unit for the purpose of taking thermograms. The strap and attached case, which can be seen hanging from the right shoulder, carry the battery pack that supplies DC power to the Thermovision system. Quite obviously, the AGA Thermovision system is very portable.

Fig. 3 shows a "heat picture" or thermogram taken from inside a newly constructed building. The black areas show the presence of cold air leaking into a room of the building at the junction of two exterior walls and through the attic space. These dark areas indicate that a great amount of heat is being wasted due to sloppy installation of insulation in the walls and in the attic space. If corrective action were taken before completion of the building, it is obvious that a significant amount of money would be saved over the lifetime of the building.

Fig. 4 shows a thermogram of the exterior of a house. The large white area on the left side of the roof reveals heat escaping through an uninsulated cathedral ceiling, while the darker part of the roof (right side) shows relatively little heat loss due to a well-insulated dropped ceiling. In this instance the AGA Thermovision reveals that, rather than an inferior insulation job, the basic architecture of a building is responsible for considerable heat loss.

3

HOW TO USE THERMOGRAPHY FOR A HEAT LOSS INSPECTION SERVICE

The previous chapter dealt with a description of the technology referred to as infrared thermography. This chapter is concerned with the application of this technology to the location of heat leaks into or out of a building, and the many advantages this technology has for becoming, in effect, the only tool that can supply both an immediate and a long-term solution to the energy crisis.

The basic operation and function of an infrared camera system were explained in chapter 2. In order to perform heat loss inspections, a system must have certain attributes. The AGA Thermovision infrared camera system has all the essential elements for making it the most practical system for use by a heat loss inspection service. For instance, it is extremely lightweight, the total system weighing only about twenty-one pounds, including the battery pack, which can provide continuous operating power for three to four hours before requiring recharging. Also, as is discussed in more detail in the following chapter, the AGA Thermovision system is extremely simple to operate.

Having such a system at your disposal, this is how you, operating as a heat loss inspection service, would proceed to make an inspection of a building. The first decision concerns the time of the heating season to begin your inspection service. This, of course, will vary with geographic location; however, the general rule of thumb is that you want to make your inspection during the times of maximum temperature difference between inside and outside. Therefore, ideally, the colder the outside air, and the warmer it is inside, the easier it will be to obtain good quality thermograms showing locations of potentially large heat leaks.

The preceding information has to be tempered by certain considerations. First, you cannot, and should not, wait until the middle of winter to begin your inspection service because businessmen and homeowners will probably want to have the opportunity to take corrective action immediately upon learning where the most serious heat leaks are occurring. If you own a building supply house in addition to the heat loss inspection service, you probably will be eager to sell the customers the insulation and possibly the services which they require to stop the heat loss. Therefore, after signing a contract with a customer to provide the heat loss inspection, you should find out when he turns on his heating system and the approximate temperature at which he maintains the building. When this temperature is at least 30° F. different from the outside temperature, you can initiate your heat loss inspection of buildings. If you attempt to start when the temperature difference between inside and outside is much less than 30° F., it will be difficult, if not impossible, to obtain satisfactory results.

The next decision concerns the appropriate time of the day or night during which you should make the heat loss inspection, and whether you should be inspecting from the inside or the outside of the building. This decision will be dictated largely by the size and type of building which you have been contracted to survey. If it is a multi-story building, or if it is only one floor but covers a rather large expanse, then it would be more prudent to survey the walls from the outside and the roof and floor from the inside. The reasoning for this is simple. Since heat can be lost only from the roof, floor, and four walls of a building, it is easier to make the wall inspection from the outside where you can observe large areas of it at one time, rather than working inside where you would have to be moving from one room to another in order to be able to get the complete picture of the heat loss situation. In some cases you may not be able to get enough distance between you and the building to observe the heat loss (e.g., buildings too close together, water on one side of the building, building located on steep hill, trees very close to the building, etc.). In those cases, of course, you will have to work on the inside. Also, if the building is multi-story or has a flat roof, the roof cannot be observed unless you are either on top of it, or are in or on an adjacent building. It would, therefore, be best to work from the top floor of the multi-story building and to inspect the roof from the inside, where pointing the infrared camera at the ceiling would indicate where cold air was coming in, and, in effect, give you the same results as working from the outside.

Making a heat loss inspection from the outside has one slight disadvantage. Because an infrared camera system will detect heat without discriminating what generates that heat, it will be sensitive to the heat from the sun as well as to that generated inside the building. Therefore, even if the sun shines for only a few minutes on a wall or roof of a building, so much heat will be absorbed by that wall or roof and reradiated that it will mask completely the heat generated inside the building. For this reason you should make your heat loss inspection at times of minimum direct solar heating of the building walls. If the inspection is made in late afternoon or early evening, it should begin with the north and east walls and end with the west and south walls in order to give the west and south walls time to cool down, especially if the day has been bright and sunny. To business and industrial clients these working hours should not prove to be bothersome because there is less commotion and interference after everyone has left for the day; however, many homeowners may not want to be bothered during evening hours, in which case you could do the inspection during the day from the inside of the house.

As was mentioned previously, working from the inside of a building has the disadvantage of requiring more time, since you will have to enter each room in order to observe the outside wall in that room; but, since houses usually are significantly smaller than business or industrial establishments, the extra time required for making the inspection from the inside should not overburden your work schedule. You will have the added advantage of not having to wait for the walls of the house to cool down because the sun has been shining on them. You will also probably find in many instances that the homeowner is so interested in what you are doing and how you are doing it that he may make an appointment with you for the evening hours to have his house inspected.

Subsequent to signing a contract with an industrial or commercial client, you should be prepared to visit the building prior to making the heat loss inspection. You should have with you a camera (preferably a Polaroid) and a note pad. The camera will be used to take photographs of the walls from both the inside and the outside of the building. On the note pad you must record the extremely important information about the materials with which the building is constructed. For instance, if you are going to inspect the inside of the building for heat losses, you will want to make notations regarding the type of interior finish and color of paint used on the ceiling and the exterior wall of each room (e.g., plaster, paneling, Sheetrock, wallpaper). Also, you will need to observe such things as the color and type of trim around doors and windows. If you will be making an inspection of the outside of the building, you will need to make note of the exterior construction materials (wood, metal, brick, stone, etc.), and the color of paint or stain on the building and on the trim around windows and doors. You will need to know the location of such things as vents and other objects that might appear in the thermograms. It will soon be obvious that the more familiar you are with the construction aspects of each building and the more notes and photographs you have, the easier it will be to analyze the thermograms for heat loss. Notes alone probably will have to suffice as information when inspecting homes, since some homeowners may, understandably, not wish to have photographs taken of their houses, especially of the interior.

After you have obtained thermograms of the building, you will compare them with your photographs and/or notes. Such a comparison is very important, for each and every building, regardless of size or shape, is constructed of a number of different materials (e.g., a wood-frame building on concrete foundation, with window trim painted a contrasting color), and therefore will appear in a range of different brightnesses on a thermogram. Each brightness corresponds to a different construction material. These different brightnesses are not due to real heat loss, but rather are referred to as apparent heat losses because the building appears to be losing heat at these locations of extreme brightness. The comparison of the thermogram with the photograph will enable you to determine whether a particular brightness on the thermogram is due to a real heat loss from insufficient or nonexistent insulation, or whether it is an apparent heat loss, i.e., the construction material in this area is different from the material used on another part of the building. The analysis of the thermograms will be the most difficult and time consuming part of the heat loss inspection and should always be done in conjunction with the regular photographs of the building to avoid making wrong judgments concerning the location of potential heat loss.

The time required for making the actual heat loss inspection will, of course, depend on the size of the building and the amount of information required by the client. Most homeowners and many businessmen will require only that you document all locations of heat loss. Others, in addition, will probably want your recommendations on what should be done to correct the situation. This is where a lot of experience in construction and insulating practices will be very valuable and also where good common sense can lead to ingenious and inexpensive solutions. Some businessmen, of course, will require additional information, such as quantity of heat being lost, to assist them in making important decisions regarding future upgrading of their facilities. Supplying this kind of information generally will require twice as much time as providing the client with simple locations of heat loss.

When calculating the amount of time required for the actual inspection, keep in mind that an average-size house will take approximately two hours, and larger buildings will take proportionately longer until you have learned exactly what has to be done and how to organize the survey. That is why, during the initial contact with your client, you should determine and, if possible, have in writing the information your client will require. In addition, you should observe very carefully all the construction details of the client's building. This will help immensely when analyzing the thermograms to determine the locations of heat losses.

As was mentioned in the previous chapter, the thermogram that shows the heat loss information is simply a Polaroid photograph of the thermal picture displayed on the TV screen of the AGA Thermovision system. The important thing to remember here is that you must label these thermograms immediately after taking them, so that you will know later if it is a thermogram of the west wall of the building, the south wall, or whatever.

Initially, the most time-consuming and most difficult part of the heat loss inspection service is the accurate analysis of the thermograms so you will be able to tell your client where heat loss is occurring and how much heat is being wasted. Immediately after inspecting the building it will not be possible, at least initially, to give your client all the details of where insulation is missing or insufficient. You will need to study all the thermograms thoroughly and make comparisons with the regular photographs to be absolutely sure that you have correctly identified locations of actual heat loss.

The fact that you will not be able, initially, to provide a complete analysis of your client's heat loss problems immediately upon completing the inspection should not discourage you. When you take a car, television set, or any appliance in for repair, frequently immediate analysis cannot be made of the problem, but must await the outcome of certain troubleshooting procedures. As you gain more and more experience in heat loss inspection, analysis time will decrease; and eventually, after having become familiar with hundreds, even thousands, of thermograms of certain locations of heat loss you will be able to immediately outline the solution for your client.

In general, you will experience few problems when making the heat loss survey of any building. The AGA Thermovision infrared camera system is small, lightweight, and easy to operate. The complete system, including battery pack, can be carried by one person by means of special harness that is sold with the system. As an alternative means of mobility, the camera and TV unit can be mounted on a tripod (sold with the system), which can be folded and carried from place to place without removing either component of the system. The battery pack, in either case, is conveniently carried over the shoulder when AC power is not available. The author prefers operating with a tripod because it is a lot easier to use the Polaroid camera that is attached to the TV unit when taking thermograms.

A few points should be emphasized. You will have to adjust the lens for the best focus on the infrared camera, just as you would normally do on any camera. This is easily done by watching the picture on the TV unit while adjusting the lens. The best focal setting for the lens, of course, will depend on where you are standing with respect to the building that you are inspecting. The ideal situation is to be close enough to the building to observe heat leaks occurring in the relatively small locations (around vents, windows, and doors), but yet far enough away so that you will not have to take too many thermograms in order to document the heat losses from the side of a building. (Unfortunately, when conducting a heat loss survey from inside a building, the distance away from a particular wall from which you can take a thermogram will be restricted by the size of the room and/or objects in the room.) Do not let the knowledge that you are working with an infrared camera intimidate you. That it is sensitive to heat rather than light does not alter the fact that it is still a camera; therefore, if you are familiar with the operation of a standard camera, with respect to such things as focal length and F number, you will have no problems with the infrared camera. You will have to be careful not to tip the infrared camera up or down too much because the coolant will spill out of the container in the camera. This is not a major problem, though, since the container holds only about one-fourth pint of coolant.

The coolant has a lifetime of two to four hours before evaporating, so you will have to take a few minutes out of your schedule every few hours to refill the container with coolant.

Almost everything that has been said about the procedures for locating heat loss in a building also applies to locating heat gain. In locating heat leaks on a thermogram, you look for bright areas (hot spots) if the thermogram is taken from the outside of a building. You look for dark areas if the thermogram is taken from the inside. The situation is just the opposite for locating heat gain in an air-conditioned building. You look for dark areas on the thermogram taken from the outside and for light areas on the thermogram taken from the inside. The major limitation in using thermography for locating heat gain in an air-conditioned building is that the minimum temperature difference of 30° F. (mentioned earlier in this chapter) that should exist between the inside and the outside of a building in order to obtain good quality thermograms will not occur until an outside temperature of 95° F. is attained. This is because most air-conditioned buildings do not, or cannot, maintain an inside temperature below about 65° F. for a range of outside temperatures between 85° F. and a maximum of 110° F. This would appear to create a severe limitation on the use of thermography for locating heat gain; however, this is not true for two reasons. As opposed to surveying for heat leaks during the cold weather when you do not want the sun to be shining on the building, during the warm weather you will want to pinpoint locations of heat gain during periods of greatest heat load, i.e., when the sun is shining directly on the wall. You will have to make your inspection from inside the building under this condition. Second, in most climates in which there are not a great number of days that the outside temperature reaches 95° F. it is probable that there will be a significant number of days when the outside temperature will be at least 30° F. below the inside temperature during the cold weather. This indicates that the building survey could be made during the cold period with no problem, and the results would be applicable for the complete range of temperatures.

The reader may be asking at this point if the use of thermography is the best or even the only way in which a heat loss inspection service can be implemented. The answer is an unqualified "yes." Until the development of the infrared camera system, no technology existed that would permit the rapid inspection of large buildings for heat leaks. It should be emphasized to the reader that since the infrared camera system displays on the TV screen heat losses as they are occurring while under such observation, any corrective action which could be taken at the moment (e.g., plugging a hole with insulation) would generate a change in the heat picture. The change will be observed immediately. Also, since the infrared camera system does not need to contact the building surface in order to detect the heat coming through the wall, an inspection of a large building can be done rather quickly.

Providing a follow-up inspection for the customer subsequent to whatever corrective action has been taken to prevent the heat loss (or gain) identified in thermograms of his building is a good way to sell a heat loss inspection service. Thermograms can again be taken of those locations identified as points of heat loss (or gain) during the initial inspection with the infrared camera system and the customer can be supplied with proof that proper insulation, or remedy, was or was not provided by the contractor responsible for correcting the problem.

4

HOW TO PROFIT FROM BUILDING HEAT LOSS INSPECTION

In this chapter the type of businessman best suited to go into the heat loss inspection business will be discussed, as well as the attitudes of the vast number of potential customers who desperately need this kind of information in order to effect a drastic reduction in their energy costs.

If, after reading chapter 2, you feel that the AGA Thermovision infrared camera system is too complex to operate because you have no background in electro-optical engineering, you are mistaken. The AGA Thermovision system has been designed so that, after only a few hours of training and working with the system, an individual with absolutely no knowledge of electro-optical engineering can operate it. The author has personally trained a number of individuals with little or no technical background in the operation of the system, and after a few hours they were able to obtain thermograms of excellent quality. The most important characteristics an operator of the AGA system can possess are common sense and good judgment.

Do not consider for one moment that you are limited in your solutions when trying to prevent excessive heat losses from homes or other buildings. The author's own electrically heated home on a windy mountain in Vermont is an example of alternative solutions. There are no windbreaks for at least 100 yards off the north-northwest exposure of the house. The planting of trees would offer a solution only after a great number of years, because of the growing time needed to compensate for the fact that they would have to be planted on land sloping rapidly away from the building. Trees, unfortunately, would also interfere with one of the major advantages of the property—a desirable view. Having access to many thermograms which showed the dramatic loss of heat from buildings subject to cold winds, the author decided to build a deck with a solid windbreak along the edge of it on the exposed side of the house. The windbreak was constructed of 1/4 inch plywood 3 1/2 feet high. The reduction in usage of electricity as a result of this effort was approximately twenty-five percent. This was determined by comparison of electric bills for appropriate periods of the previous year. This dramatic reduction was accomplished with an investment of about $200, including materials and labor. The windbreak paid for itself the first winter.

As if to offer a means of testing the validity of the evidence, late in the winter the windbreak was sheared off the deck by falling chunks of ice and snow that slid off the roof after a few days of freezing rain. (Theoretically, this should not happen with a high-pitched asphalt shingle roof. So much for theory.) The windbreak was not retrieved but remained in the snowbank around the house. This was a mistake. The next electric bill indicated that usage was up twenty percent. After that, the windbreak was retrieved and simply placed against the outside of the house. The next electric bill was reduced by that same twenty percent. The percentage of reduction in use of electricity is valid for Vermont in late winter months since these months are still winter months with cold wind. A better controlled experiment could not be desired by any scientist.

Do not be fooled by thinking that there will be little or no business in the heat loss survey field in the new-home market. The author contracted the construction of the shell of his home to a reputable building contractor in the area. Yet such omissions as no insulation around the vent for the cooktop and none between door casings and walls indicate that new buildings are being

adjacent walls? There are only two ways to determine whether or not valuable heat is being lost in these two locations. The homeowner can either remove all the casings and the Sheetrock behind them (and run the risk of making a mess of the job), or he can call for the services of a heat loss survey contractor who can tell the homeowner in a few seconds if he is losing valuable heat at those locations. From there, of course, in a matter of only two or three hours (depending on the size and architectural complexity of the building), a complete survey can be made of the house and the homeowner can be made aware of where the major heat losses are occurring.

One of the major attractions of using thermography to find heat losses in business and residential buildings is the potential for increased sales of all types of insulation, once the locations of heat losses are determined and the situation analyzed. Heretofore the average homeowner, businessman, and industrialist have relied upon an abundance of cheap energy which supposedly eliminated any need to worry about whether or not heat was escaping because of faulty or nonexistent insulation. However, today the situation is entirely different, and building supply houses selling all forms of insulation (glass fiber, reflective aluminum foil, blown cellulose fiber, poured vermiculite and perlite, polystyrene and polyurethane foams) will experience a booming business when millions of businessmen and homeowners discover through thermography that their buildings are sadly lacking in thermal integrity.

From the author's point of view, after many, many hours of thought concerning the problem, individuals who are at present in the building supply business or are considering opening such a business will find the fastest return on their investment capital. The reason for this is the ability to provide not only the heat loss inspection service but also the needed insulation and its installation. It will be convenient for the homeowner and businessman to buy the product as well as the service.

However, do not be discouraged if you are not already in that business, because, with the potentially huge volume of all types of insulation that will be sold in the future as a direct result of building heat loss inspections, possibly you will be able to enter into a business deal with your local building supply house. You have not reached the end of your profit potential with thermography alone. If you have the expertise, the tools, and the materials you can offer the customer the complete job from beginning to end, i.e., from survey to final installation of insulation where it is needed.

The potential for sale of insulation is fortunately only half the story. With the rapid expansion of the use of thermography as a tool to locate costly heat losses, another huge market is looming ahead. Energy conservation consulting in the business sector will blossom into a multi-million dollar market simply because harried business executives and industrialists are extremely eager to cut needless costs in order to be able to report higher profits to their stockholders. The development of this new technology will enable facility management consultants to expand their business activities to include heat loss inspection services. As a matter of fact, this aspect of facility management may rapidly become the most important and profitable function with regard to reducing energy costs in buildings.

The reader probably will ask what type of building can be inspected for heat loss or heat gain. The answer is simple: Any building that is heated or air conditioned. This includes one-story buildings, multi-story buildings, buildings of conventional construction, buildings of contemporary construction, new houses, old houses, brick houses, frame houses, log houses, prefabricated houses, office buildings, apartment buildings, metal buildings, warehouses, stores, factories, schools, churches, hospitals, laboratories, museums, mobile homes, etc. The list is endless. In fact, there is absolutely no type of building that cannot be inspected for heat loss or heat gain.

At what locations on building surfaces can one expect to find heat leaks or excessive heat losses? This will depend, of course, on whether it is a large industrial building or just an average-size home. The following are a number of locations that should be inspected for heat leaks: (1) the corner of a building or the joint of two prefabricated sections of a building, (2) the joint between roof flashing and chimney, (3) the joint between window and wall and door and wall, (4) the joint between basement wall and floor, (5) around loose-fitting or open roof vents, (6) through eave and gable vents, (7) around studs and joists that have insufficient or loose-fitting insulating, or no insulation at all, (8) around improperly installed window air conditioners, (9) around the glass in windows with brittle caulking material, (10) at the place where radiators are mounted on or very close to the wall, and (11) through single pane windows. The reader can easily see that heat leaks or excessive losses can occur at many places, and it does not take much effort to pinpoint these locations with an infrared camera. That is why, in the final analysis, the heat loss inspection service will be the moneymaker of the future.

The time is ripe to introduce the concept of heat loss inspection services to the general public. Everywhere that the author has traveled and given seminars, a great deal of interest has been generated because homeowners and businessmen alike simply can no longer afford to pay higher and higher energy costs without soon going bankrupt. It is as simple as that.

5

LIMITATIONS

The author would be less than candid if he did not discuss some of the difficulties involved in the use of thermography for a heat loss inspection service; therefore, this chapter will be devoted to a discussion of a few limitations which every businessman must consider before embarking upon this venture.

The first consideration is that the AGA Thermovision system is a delicate piece of electro-optical equipment that requires the same careful handling accorded an expensive camera or binoculars. This does not mean that undue care is necessary, for the author has used the system in surveying the thermal integrity of a cooling system for a nuclear reactor, subjecting the AGA system to very high humidity with temperatures in excess of 80° F. The system was also carried up and down twenty-five foot ladders and then taken into a chamber cooled to 15° F. However, if you happen to drop either the infrared camera or the television unit, it will more than likely be damaged enough to put you out of business temporarily.

The unit inside the camera which senses the heat and converts it into an electrical signal requires cooling with liquid nitrogen. This simply means that after every two to four hours of operation (or before, if you accidentally spill coolant) the container in the camera must be filled with about 1/4 pint of the coolant. Since the coolant is at an extremely cold temperature (—196° C.), some care must be exercised in not allowing it to splash into anyone's eyes. However, simply by the exercise of good judgment and common sense, serious problems can easily be avoided.

Turning now to the interpretation of the thermograms, there are two problems that must be reconciled if you as a businessman are going to be selling a heat loss inspection service to the public. The first problem that you will encounter is that of deciding whether or not the brightness that you see in each thermogram is apparent or real heat loss (see chapter 3). In many thermograms the answer will be obvious; however, in many others it will not be. There is no easy solution to this problem, since apparent heat losses can be attributed to a variety of circumstances. But, as you gain more experience in making heat loss inspections, interpretation of thermograms will consume less and less time and become less difficult. To assist you with problems of interpretation, the AGA Corporation can recommend a consulting engineer (see the last paragraph of chapter 6).

Many homeowners and businessmen will only need to see thermograms of their buildings to be convinced that insulation is missing or insufficient, and they will be ready to purchase your materials and services (if you are already in the building supply business). In many business situations, however, large amounts of money will be at stake in the decision either to invest in new facilities or to upgrade present facilities, and executives will need to have quantitative data on which to base their final decision. This will necessitate a very careful inspection of the facilities to determine at what locations heat is being wasted. Once this is determined, calculations must be made to compare the actual heat being wasted with that of the design heat loss of the facilities involved. This comparison will then enable the business executive to make a judgment based on a fairly accurate projection of energy costs which he could not obtain in any other manner.

It is obvious that this type of heat loss inspection service requires the assistance of a consulting engineer to aid in calculating the design heat loss and in making the necessary comparisons with the actual heat loss. The last paragraph of chapter 6 gives specific information.

No new venture with so much profit potential can be expected to develop without any problems whatsoever. One need only consider the development of the laser during the past fifteen years to see that many problems had to be overcome before practical application could be realized, but laser systems are now being employed rather routinely from assembly lines in factories to checkout counters in supermarkets. Thermography, however, is already being employed in a practical manner by some members of the medical profession. It is being used for the early detection of breast cancer. It also reveals the location of inflammations such as phlebitis. Increase in skin temperature apparently accompanies many pathological conditions, and thermography appears to be a significant tool in locating these conditions. In addition to its use by the medical profession, thermography is being used widely in Sweden, Canada, Great Britain, and the United States for preventive maintenance by the electrical power industry. The inspection of transmission lines by helicopter and ground patrols using infrared thermography to find overheated circuit breakers, transformers, busbar joints, etc., has proved to be far superior to any other method used. More important, maintenance and inspection costs have been dramatically reduced by the use of this method.

Some mention must be made of the capital investment required for the AGA Thermovision infrared camera system. The cost of the basic system is approximately $40,000. While this certainly is a considerable investment, it should be emphasized that very few businesses can be started with less capital and yet have the potential for such enormous profits.

None of the limitations mentioned in this chapter should discourage an individual from going into the heat loss inspection business. Because of the continuing energy crisis and the high cost of that energy, the heat loss inspection business will become the most sought after service in this country.

6

BUILDING HEAT LOSS INSPECTION AND THE FUTURE

This book has attempted to show the readers that, because of the energy crisis and the ever-increasing cost of energy, a means is needed to show consumers, both businessmen and homeowners, where and how much energy is being wasted every day by insufficient or a complete lack of insulation in the buildings and houses they occupy. For example, if the annual heating cost (homes, factories, office buildings, etc.) of New England alone were considered, it would amount to a staggering $2.3 billion. It is the opinion of the author that every house and other type of building in New England and elsewhere is wasting at least ten percent of its heating fuel due to inadequate insulation. In many instances in excess of twenty percent of building heat is being wasted. If corrective actions were taken in New England alone to stop this indiscriminate heat waste, the annual savings could be anywhere from $115 million to $230 million. The result of this action would be to drastically reduce United States oil imports from the Middle East, while saving the individual homeowner and businessman thousands, perhaps millions, of dollars over his lifetime.

The reader may ask why any homeowner or businessman needs a heat loss inspection. Cannot a homeowner, for example, simply put additional insulation in the attic, or have some blown into the wall, or just put up evacuated insulated glass or storm windows? There are several answers to these questions. One of these can be given through the experience of a good friend. This friend purchased a large old New England home two years ago. The previous owner told the new owner that several years before the time of the sale of the house he had contracted for insulation to be blown into the walls. The job was done, the insulation contractor got his money, and the owner apparently was satisfied. One day the new owner started some renovations. After he had removed the plaster from the wall, he noticed paper bags fastened between the studs. Upon further investigation he found that these bags were filled with sawdust, an early attempt at insulation. Some of the blown insulation was visible, but not very much. Since the previous owner did not know that he had some kind of insulation already in the walls, he thought that by having insulation blown in he would be taking the most cost-effective approach for reducing fuel consumption. In fact, he probably took the least cost-effective approach, since there was not enough room in the walls to have a really significant amount of insulation blown into the existing space. It is also a well known fact that, after several years, blown insulation tends to settle, thus diminishing its insulation potential. Had a heat loss inspection service been available, the previous owner of this house would have found out that some sort of insulation was in the walls, even if he did not know what kind. Rather than contracting for blown insulation he might have chosen, for example, to put an insulated siding on the house. In the long term, that would have been more cost-effective than the other approach.

The preceding story illustrates an important point that must be emphasized and re-emphasized to both businessmen and homeowners. The heat loss inspection service serves as a problem identifier, showing where the most serious heat losses are occurring so that the consumer (1) will be aware of where and how much heat is being lost and (2) will be able to take the appropriate corrective action rather than just guessing where and how much insulation will be needed. Even after he has contracted for the insulating services, his need for the heat loss inspection service has not ended. What is the reason for this? Unfortunately, the fact that an insulation job has been contracted for and supposedly completed does not guarantee the consumer that it was done properly. A large quantity of insulation piled deep in the attic space or anywhere else does not do much good if it is not sealing the cold air out of the building. This insulation must be tucked into small cracks everywhere that these cracks exist. If it is not, then the consumer is simply wasting his money by having the job done. A heat loss inspection with an infrared camera system can determine immediately whether or not the installation was proper.

What about new houses? Are they not being insulated better than old houses were? The answer is: Absolutely not! In an earlier chapter the author related the dismaying number of places in his home that have been found to have no insulation. He has seen thermograms of other newly constructed houses with big gaps where insulation was left out. Make no mistake about the fact that this kind of sloppy insulation installation is costing each consumer many hundreds of dollars annually. The author's heating bill for the first winter in a new house was twice what it should have been because insulation had been left out in various places. Remember, no matter how much research has been done in the past and how many building code regulations have been legislated, in the final analysis the cost of heating fuel for a newly constructed building or a renovated one will be proportionate to the quality of the insulation job.

The AGA Thermovision infrared camera system described in this book is the only truly portable battery-operated system available for making heat loss inspections of buildings. The AGA Corporation, located at 550 County Avenue in Secaucus, New Jersey, 07094, is preparing a manual of guidelines for thermo gram interpretation which will be available in the near future. They also may be contacted to provide you with a consulting service to assist you in establishing a heat loss inspection service.